Preface

This reference was written to help new students of anatomy cope with the seemingly endless amount of material in a human anatomy course. Remembering all that is presented can be extremely difficult, especially as one struggles to understand rather than just memorize. We hope that we can help students accomplish both.

We feel that the finest way to learn anatomy, is of course, to dissect, dissect, dissect, and then dissect even more. There is no substitution for the hands on experience, which creates the foundation of one's anatomical knowledge, however this reference has been designed to aid in the many endeavors a student will encounter in human anatomy.

"That which is learned "hands on" is not readily forgotten"

Weather used to prepare for the *Medical National Board Exam* or to test in an undergraduate anatomy course, <u>The Flash Anatomy Board Review</u> was written to help students truly learn and remember human anatomy. As a recognized publisher in the health allied field, we are most concerned with how this reference helps the student, therefore suggestions are welcomed and appreciated.

About the Author...

The author of
Flash Anatomy, The Anatomy Board Review, Reed S. Oxman, M.D., F.A.C.E.P., is a graduate from U.C.L.A. Medical School, and is a graduate of the Emergency Medicine Residency Program at the Medical College of Pennsylvania. He holds an undergraduate degree in Anatomy & Physiology from the University of California at Berkeley. He received Orthopedic Surgical training from the Hospital of the University of Pennsylvania. He is also *Board Certified in Emergency Medicine* from the University of Pennsylvania.

TABLE OF CONTENTS

Chapter 1 - Upper Limb 1

Chapter 2 - The Thorax 16

Chapter 3 - The Abdomen 30

Chapter 4 - Pelvis & Perineum 42

Chapter 5 - Lower Limb 54

Chapter 6 - Back 69

Chapter 7 - Head & Neck 73

Chapter 8 - Autonomic Nervous System 107

Chapter 9 - Joints 111

Chapter 10 - Developmental Anatomy 114

Chapter 1

What is Hilton's Law?

The axillary artery can be considered to be divided into three parts by the pectoralis minor muscle. What are the arterial branches of each part?

What ventral primary rami form the origin of the brachial plexus?

The posterior cord of the brachial plexus gives off three branches in addition to the axillary and radial nerves. Name them.

What are the muscles they innervate.

Name the structure(s) that pierce the clavipectoral fascia.

Card # 1

The nerve trunk supplying a muscle also tends to give off a branch of supply to the joint which it moves and another branch to the skin over that joint.

1st: supreme thoracic a.
2nd: thoracoacromial a.
    lateral thoracic a.
3rd: subscapular a.
    posterior humeral circumflex a.
    anterior circumflex a.

C5 to TI

---

upper subscapular n.
thoracodorsal n.
lower subscapular n.,

---

upper part of subscapularis m.
latissimus dorsi m. lower part of subscapularis m.
and teres major m.

---

cephalic v.
lateral pectoral n.
thoracoacromial a.
lymphatics

Name the muscle(s) enclosed by the clavipectoral fascia.

Name the rotator cuff muscles.

Name the bones that help form the apex of the axilla.

Describe the lymphatic drainage of the mammary gland.

Describe the boundaries of the deltopectoral triangle.

From what direction is the axilla most vulnerable to injury?

Card # 2

| pectoralis minor
subclavius |

| infraspinatus
subscapularis
supraspinatus
teres minor |

| clavicle
first rib
scapula |

| medially: to opposite breast
laterally: to axillary nodes |

| clavicle
deltoid m.
pectoralis major m. |

| anterior |

The manubriusternal joint marks which rib level?

The medial pectoral nerve innervates what muscle(s)?

Which division(s) of what trunk(s) form the medial cord of the brachial plexus?

Which division(s) of what trunk(s) form the lateral cord of the brachial plexus?

Which division(s) of what trunk(s) form the posterior cord of the brachial plexus?

Name the origin(s), insertion(s), and innervation of the deltoid muscle.

Card # 3

2

pectoralis major and minor

anterior division of the lower trunk

anterior division of the upper and middle trunks

posterior division of the upper, middle, and lower trunks

o: lateral 1/3 of the clavicle
acromion process, crest of scapular spine
i: deltoid tuberosity of humerus
n: axillary n.

The clinical condition known as "winged scapula" (paralysis of serratus anterior muscle) results from the disturbance of what nerve?

Name the origin, insertion, and function of teres minor.

What blood vessel passes over the suprascapular ligament? What nerve passes beneath the ligament?

What nerve innervates the trapezius muscle?

Name the origin, insertion, innervation, and function of the infraspinatus muscle.

Name the arteries that take part in the extensive anastomoses in the scapular region.

Card # 4

long thoracic

o: lateral border of scapula
i: lowest facet of the greater tubercle of the humerus
f: laterally rotates and adducts the arm

suprascapular a.

suprascapular n.

Spinal Accessory (XI)

o: infraspinous fossa
i: middle facet of greater tubercle
n: suprascapular n. (C5, 6)
f: laterally rotates the arm

intercostal
subscapular
suprascapular
transverse cervical

Name the origins of the trapezius muscle.

What are the three major arterial branches of the thyro-cervical trunk?

Name the origin, insertion, innervation, and function of latissinmus dorsi muscle.

√

What are the five terminal nerves of the brachial plexus?

√

Name the origin, insertion, and function of the subscapularis muscle.

√

What two major nerves traverse the arm without giving off muscular branches?

Card # 5

superior nuchal line
ligamentum nuchae
spinous processes and supraspinous
ligaments of vertebrae TI - T12

---

suprascapular a.
transverse cervical a.
inferior thyroid a.

---

o: lumbar aponeurosis
i: bicipital groove of the humerus
n: thoracodorsal nerve
f: medial rotation, adduction,
and extension of the humerus

---

axillary
median
musculocutaneous
radial
ulnar

---

o: subscapular fossa
i: lesser tubercle of the humerus
f: medial rotation of the arm

---

median
ulnar

Name the innervation of the flexor and extensor brachial muscle groups.

Name the muscle(s) comprising the flexor and extensor brachial muscle groups.

What is the common insertion of the triceps muscle?

Name the groove between the medial and lateral heads of the triceps. Name the contents of this groove.

Name the origin(s) and insertion(s) of the biceps brachii.

What nerve provides cutaneous innervation to the posterior arm and forearm?

Card # 6

flexor: musculocutaneous n.
extensor: radial n.

---

flexor: biceps
brachialis
coracobrachialis
extensor: triceps

---

olecranon process of the ulna

---

spiral groove

profunda brachii a.
radial n.

---

o: -short head — coracoid process
long head — supraglenoid tubercle
i: radial tuberosity

---

radial

The axillary nerve innervates what muscle(s)?

What muscle(s) flex the elbow joint?

What muscle(s) extend the elbow joint?

Name the contents of the cubital fossa from lateral to medial.

What are the boundaries of the cubital fossa?

Name of the muscle(s) originating from the medial epicondyle of the humerus.

Card # 7

- deltoid
- teres minor

- brachialis
- biceps brachii
- brachioradialis
- pronator teres

- triceps
- anconeus

- radial n.
- deep branch of radial n. (Post interosseous n.)
- biceps tendon, brachial a.
- median n.

- brachioradialis m.
- pronator teres m.
- the line between humeral epicondyles

- pronator teres
- flexor carpi radialis
- palmaris longus
- flexor carpi ulnaris
- flexor digitorum superficialis

Name the major flexor of the elbow joint.

Name the muscle(s) originating on the lateral epicondyle of the humerus.

What ligament creates a gliding joint between the humerus and radius in addition to a pivot joint between the ulnar and radius?

Though the brachioradialis muscle is a flexor at the elbow, it is innervated by what nerve?

What extensor muscles originate from the supracondylar ridge of the humerus?

What type of circulation prevents the hazardous effects of brachial artery occlusion at the elbow joint?

Card # 8

brachialis, m.

extensor carpi radialis brevis
extensor digitorum
extensor digiti minimi
extensor carpi ulnaris
anconeus

annular

radial

brachioradialis
extensor carpi radialis longus

collateral

Name the muscle(s) in the superficial group of forearm flexors.

What muscle(s) pronate the forearm?

Name the muscle(s) in the intermediate group of forearm flexors.

Name the muscle(s) in the deep group of forearm flexors

The deep radial nerve pierces what muscle to supply the posterior forearm muscles?

Supination of the forearm is performed by what muscle(s)?

Card # 9

- pronator teres
- flexor carpi radialis
- palmaris longus
- flexor carpi ulnaris

- pronator teres
- pronator quadratus

- flexor digitorum superficialis

- pronator quadratus
- flexor pollicus longus
- flexor digitorum profundus

- supinator

- biceps brachii (primarily)
- supinator
- brachioradialis

What bony structure acts to deflect the direction of the extensor pollicis longus tendon?

In the forearm, the median nerve runs in a plane between what two muscles?

What artery passes through the anatomical snuff box?

Name the muscles of the deep group of the extensor forearm region.

What is the common origin of most of the extensor muscles of the forearm?

What is the innervation of the forearm flexor muscles? Extensor muscles?

Card # 10

| dorsal radial tubercle |

| flexor digitorum superficialis |
| flexor digitorum profundus |

| radial |

| abductor pollicis longus |
| extensor pollicis brevis |
| extensor pollicis longus |
| extensor indicis |
| supinator |

| lateral epicondyle of the humerus |

| flexors: ulnar n. and median n. |
| extensors: radial n. |

Name the muscle(s) of the forearm innervated by the ulnar nerve.

Name the forearm artery that may be mistaken for a vein.

Name the syndrome associated with compression of the median nerve at the wrist joint.

Which branch of the median nerve does not traverse the carpal tunnel?

Name the tendonous borders of the anatomical snuff box (lateral to medial).

What are the actions of the anconeus muscle?

Card # 11

flexor digitorum profundus
(one-half of the muscle)
flexor carpi ulnaris

---

ulnar

(important to note clinically because of the artery's possible superficial course)

---

carpal tunnel syndrome

---

palmar cutaneous branch

---

lateral: abductor pollicis longus and extensor pollicis brevis
medial: extensor pollicis longus

---

abduction of the ulna during pronation
extension of the elbow joint

What artery runs beside the deep radial nerve in the posterior compartment of the forearm?

What arterial source(s) provide blood to the dorsal region of the hand?

Name the bone serving as the common origin of the hypothenar muscles.
Name the muscle that inserts here.

Name the eight carpal bones.

Name the extensors of the thumb.

Name the vascular structures in the palm formed by the continuations of the ulnar and radial arteries.

Card # 12

posterior interosseous

radial
dorsal carpal branch
of the ulnar

pisiform

flexor carpi ulnaris

hamate, pisiform, triquetrum, lunate
capitate, scaphoid, trapezoid
trapezium

abductor pollicis longus m.
extensor pollicis brevis m.
extensor pollicis longus m.

superficial palmar arch
deep palmar arch

The synovial sheaths of which digits are separate from the common synovial sheath of the palm?

What nerve is responsible for wrist and thumb extension?

Blood from the hand drains into the dorsal venous arch which in turn drains into what vein laterally? Medially?

What nerve is responsible for finger abduction and adduction?

The wrist joint is capable of what movements? Not capable?

What are the origins and insertions of the four lumbrical muscles?

Card # 13

2
3
4

radial

cephalic

basilic

ulnar

abduction, adduction
circumduction, extension
flexion

not rotation

o: flexor digitorum profundus tendons
i: lateral sides of extensor expansion

Name the muscle responsible for index finger abduction? Adduction?

Name the marginal projections of the carpal bones which allow attachment of the flexor-retinaculum.

Draw the distribution of cutaneous nerves to the dorsum of the hand.

Draw the distribution of cutaneous nerves to the palm of the hand.

What muscles of the hand are innervated by the median nerve?

What muscles form the hypothenar eminence and what is their innervation?

Card # 14

- dorsal interosseous
- palmar interosseous

---

medial: pisiform and the hook of hamate
lateral: tubercles of scaphoid and trapezium

Refer to figure 1:1 A

Refer to figure 1:1 B

---

thenar group:
abductor pollicis brevis
flexor pollicis brevis
opponens pollicis
first two lumbricals

---

abductor digiti minimi
flexor digiti minimi
opponens digiti minimi

ulnar n.

Name the lumbrical muscles innervated by the ulnar nerve.

What nerve supplies the interossei muscles?

Name the clinically important branch of the median nerve associated with the thenar muscle group.

The radial artery pierces which interosseous muscle?

Name the muscles primarily responsible for the power of the pincer grip.

The common interosseous artery is a branch of what larger artery?

Card # 15

3rd and 4th lumbricals
ulnar
recurrent branch of the median n.
first dorsal interosseous
flexor pollicis longus opponens pollicis flexor digitorum profundus
ulnar

Chapter 2

What is the vertebral level of the suprasternal notch?

What prevents the upward displacement of the humerus?

Name the divisions of the mediastinum.

What is the name of the concave impression in the left lung?

The azygos vein drains into what vessel?

To what rib number does the pleura extend inferiorly at the midclavicular line? Midaxillary line? Spine?

Card # 16

T2

coraco-acromial arch

superior
inferior: anterior
middle
posterior

cardiac notch

superior vena cava

R8
R10
R12

In general, how many ribs above the most inferior extent of the pleura are the lungs?

What vertebral levels demarcate the extent of the trachea?

What vessel gives rise to the bronchial arteries?

Refer to figure 2:1

Name the origin and branches of the internal thoracic (mammary) artery.

Name the loose fold of pleura that allows for lung root movement during respiration.

Name the four divisions of the parietal pleura and their innervations.

Card # 17

	2
	C6 - T4/5
	thoracic aorta
	o: subclavian a. branches: musculophrenic a. pericardiophrenic a. superior epigastric a. upper 5 anterior intercostals
	pulmonary ligament
	cervical: segmentally supplied costal: segmentally supplied diaphragmatic: phrenic & lower 5 intercostal nerves mediastinal: phrenic n.

The bronchial veins drain into what vein(s)?

What effects are produced in the lungs by sympathetic nervous activity? Parasymphathetic nervous activity?

Describe by layers the trilaminar insertion of the pectoralis major muscle in the intertubercular groove of the humerus.

What are boundaries of the deltopectoral triangle?

What is the origin, insertion, and innervation of the pectoralis minor muscle?

What is the innervation of the pectoralis major muscle?

Card # 18

azygos
hemiazygos

---

sym: bronchodilation
vasoconstriction

para: bronchoconstriction
vasodilation
glandular secretion

---

anterior layer: clavicular fibers
intermediate layer: upper sternal fibers
posterior- layer: lower sternal fibers

---

deltoid m.
pectoralis major m.
middle 1/3 of the clavicle

---

ribs 3, 4 and 5
coracoid process
medial pectoral n.

---

medial and lateral
pectoral nerves

What are the four main branches of the thoraco-acromial artery?

Refer to figure 2:2
Name the four major structures contained within the deltopectoral triangle.

Travelling just beneath the edge of each rib are two blood vessels and a nerve. What is their order from superior to inferior?

In what direction are the fibers of the external intercostal muscles directed? Internal intercostal muscles?

The xiphisternal joint lies opposite to what vertebral level?

The sternal angle lies opposite to what vertebral level?

Card # 19

- pectoral a.
- clavicular a.
- acromial a.
- deltoid a.

- cephalic v.
- clavipectoral fascia
- lateral pectoral n.
- thoracoacromial a.

---

vein
artery
nerve

---

downward and forward

downward and backward

T9

T4/5

Name the anterior and posterior landmarks marking the imaginary plane that divides the mediastinum into superior and inferior regions?

Accidentally inhaled small objects are usually found in what primary bronchus?

What major vessels gives rise to the posterior intercostal arteries? The anterior intercostal arteries?

What gland in the thoracic cavity is significantly larger in a child than in an adult?

What nerve passes directly inferior to the ligamentum arteriosum?

Name the bony tube formed by the consecutive vertebral foramina.

Card # 20

anterior: sternal angle posterior: vertebrae T4/5
right
thoracic (descending) aorta and the superior intercostal a. internal thoracic a.
thymus
left recurrent laryngeal
vertebral canal

What are the four main parts of a rib?

Name the three parts of the sternum.

Name the fibrocartilaginous structure that contributes to the symphysis joint between vertebrae.

The mammary papilla lies at what intercostal space in the male?

The apex of the heart lies at what intercostal space?

Name a synonym for the apex of the lung.

Card # 21

	body neck head tubercle
	body manubrium xiphoid process
	intervertebral disc
	4
	5
	cupola

What ligaments secure and compartmentalize the superficial fat of the mammary gland?

What is the innervation of the platysma muscle?

What nerve(s) innervate the skin that lies over the platysma muscle?

What would dimpling of the skin of the mammary gland possibly indicate?

Name the three parts of the thoracic cavity.

Name the term used to describe the line denoting the transition from costal pleura to diaphragmatic or mediastinal pleura.

Card # 22

- suspensory ligaments of Cooper
- a branch of the facial (VII) nerve
- supraclavicular nerves
- shortening of Cooper's ligaments (as in breast cancer)
- left and right pleural sacs mediastinum
- pleural reflections

What is the function of the serous lubricating fluid found in the pleural cavities?

What are the terms used for air, blood, and lymph accumulation in the pleural cavity?

Name the procedure used to remove pathological materials from the pleural cavity.

Which intercostal spaces are usually selected for pleural tapping.

What nerve innervates the diaphragm?

What ventral primary rami contribute to the phrenic nerve?

Card # 23

to reduce friction between the parietal and visceral pleura during respiration

pneumothorax
hemothorax
chylothorax

Pleural tap (pleuracentesis)

6, 7, and 8

phrenic

C3, 4, and 5 (mainly C4)

What type of blood, venous or arterial, is carried by the pulmonary artery?

What is another name for the visceral layer of the serous pericardium?

Name the pericardial sinus on the posterior surface of the heart produced by the reflection of the serous pericardium around the large veins.

Name the pericardial sinus on the posterior surface of the heart that lies between the aorta and pulmonary veins.

What groove, running obliquely around the heart separates the atria from the ventricles?

What grooves separate the ventricles from one another?

Card # 24

- venous (deoxygenated)
- epicardium
- oblique pericardial sinus
- transverse pericardial sinus
- coronary sulcus
- anterior and posterior interventricular grooves or sulci

Name the origin of the right and left coronary arteries.

Name the three major branches of left coronary artery.

Name the two major branches of the right coronary artery.

Into what vessel do most of the cardiac veins drain?

The coronary sinus drains into what structure?

What important cardiac vein is located in the anterior interventricular sulcus?

Card # 25

- ascending aorta (just behind the aortic cusps)
- anterior interventricular a.
  circumflex a.
- inferior interventricular a.
  sinoatrial a.
  marginal a.
- coronary sinus
- right atrium
- great cardiac

What important cardiac vein is located in the inferior interventricular sulcus?

What ridge separates the anterior and posterior walls of the right atrium?

Name the characteristic muscle found in the right atrium.

What valve connects the right atrium to the right ventricle?

What is the pacemaker of the normal heart?

In what structure is the contraction signal of the heart delayed?

Card # 26

- middle cardiac
- crista terminalis
- pectinate
- tricuspid
- sinuatrial node
- atrioventricular node

What structures connect the papillary muscles of the ventricles to their respective valvular cusps?

Name the orifice of the pulmonary trunk.

Through what type of valve does blood pass on its way out of the ventricles?

What valve separates the left atrium from the left ventricle?

What characteristic invaginations are found on the walls of the ventricles?

What nerve lies anterior to the esophagus and contributes to the esophageal plexus? Posterior to the esophagus?

Card # 27

- chordee tendineae
- conus arteriosus (infundibulum)
- semilunar
- mitral
- trabeculae carneae
- left vagus
- right vagus

What major lymphatic vessel lies between the azygos vein and the descending aorta?

What major peripheral lymphoid organ lies in the superior mediastinum?

What two large veins form the superior vena cava?

Which side of the body receives a brachiocephalic artery?

At what level does the aortic arch turn downward?

The right recurrent laryngeal nerve wraps around what artery?

Card # 28

thoracic duct
thymus
right and left brachiocephalic veins
right
T 4/5 (sternal angle)
right subclavian

At what level does the trachea bifurcate?

Name the three major sympathetic nerves found in the thoracic cavity.

Name the right lung lobe that is not found in the left lung?

What two right bronchopulmonary segments are combined in the left lung?

What two bronchopulmonary segments contribute to a tongue-shaped structure in the left lung?

What arteries supply the breast?

Card # 29

T 4/5 (sternal angle)
greater splanchnic (T5 - T9) lesser splanchnic (T10 - T11) least splanchnic (T12)
middle lobe
apical and posterior
superior and inferior lingual
intercostals internal thoracic lateral thoracic thoracoacromial

What is the innervation of the pectoralis major muscle?

Chapter 3

At what verterbral levels do the following structures lie: The umbilicus? The bifurcation of the aorta? The bifurcation of the inferior vena cava?

Name the fat finger appendages found on the colon.

Name the specialized outer longitudinal muscle layer of the large intestine.

Name the origin and insertion of the rectus abdominus muscle.

What two prominent anterior abdominal wall nerves originate from LI ventral roots?

Card # 30

medial and lateral pectoral nerves
L3/4 L4 L5
appendices epiploicae
taeniae coli
o: costocartilage of ribs 5, 6, and 7 i: public tubercle
ilio-hypogastric ilio-inguinal

What major vessel gives rise to both the ovarian and testicular arteries?

Outline the walls of the inguinal canal.

What traverses the inguinal canal in the male? In the female?

Name the arteries that form the spermatic cord.

What is another name for the mouth of the omental bursa?

What two ducts unite to form the duodenal papilla?

Card # 31

	abdominal aorta (at LI)
	ventral: aponeurosis of the external oblique m. dorsal: conjoined tendon superior: arches of internal oblique and transversus abdominus m. inferior: inguinal and lacunar ligaments
	spermatic cord round ligament of the uterus
	cremasteric deferent duct testicular
	epiploic foramen
	common bile main pancreatic

Name the major openings in the diaphragm? At what spinal level is each opening found?

What structures pass through the aortic hiatus?

What three major vessels pass through the right free margin of the lesser omentum.?

Name the branches of the celiac trunk

What are the boundaries of the epiploic foramen?

What vessel receives the drainage from the left gonadal (testicular or ovarian) vein? What vessel receives the right gonadal vein?

Card # 32

- aortic hiatus T12
- esophageal hiatus T10
- vena caval foramen T8

- azygos vein
- descending aorta
- thoracic duct

- bile duct
- hepatic artery
- portal vein

- common hepatic a.
- left gastric a.
- splenic a.

- inferior vena cava
- right free margin of the lesser omentum
- first part of the duodenum
- caudate process of the liver

- left renal vein
- inferior vena cava

What are the main tributaries of the portal vein?

What muscle controls the release of bile into the duodenum?

What part of the liver is directly applied to the diaphragm?

The gall bladder drains into the bile duct through what structure?

The spleen usually lies between what rib levels?

What vessels pass through the porta hepatis?

Card # 33

- superior mesenteric v.
- splenic v.

- sphincter of Oddi

- bare area

- cystic duct

- ribs 9 - 11

- common hepatic duct
- hepatic artery
- portal vein

Name the sacculations within the colon.

What is the insertion of the diaphragm?

What two muscles form a functional unit that acts as the most powerful flexor of the thigh?

What nerve pierces the anterior surface of the psoas muscle?

Which kidney is positioned somewhat lower in the body? What is the reason for this?

What is the most posterior structure at the kidney hilum?

Card # 34

haustra

central tendon

iliacus
psoas

genitofemoral (L1, 2)

right
liver on right side

ureter

Name the dark tissue areas seen in the renal medulla.

Describe the flow of urine from the collecting duct to the bladder.

Name the layers composing the spermatic cord.

(continue to next question)

What are the major branches of the superior mesenteric artery?

What structures receive blood from the inferior mesenteric artery?

Card # 35

renal pyramids

minor calyces
major calyces
renal pelvis
ureter

external spermatic fascia
(external oblique m.)
cremasteric fascia

(internal oblique) and
(transverse abdominus m.)
internal spermatic fascia
(transversus abdominus m.)
cremaster m.

ileocolic a.
right colic a.
middle colic a.
jejunal - ileal a.

left half of transverse colon
descending colon
sigmoid colon
upper rectum

What clinical symptom may result from portal venous hypertension?

What venous plexus drains both portally?

What venous plexus drains both portally and systemically?

Heavy but usually insufficient anastomoses between the superior and inferior mesenteric arteries form what structure?

What are the major branches of the common hepatic artery?

Along which curvature of the stomach do the right and left gastroepiploic arteries course?

Card # 36

hemorrhoids esophageal varices
rectal venous plexus
rectal venous plexus
marginal artery
gastroduodenal a. hepatic a. right gastric a.
greater curvature

Which spinal nerve innervates the skin around the umbilicus?

Destruction of the inferior vena cava or the portal vein may lead to engorged superficial epigastric veins. What is the clinical name for this condition?

Name the deep layer of superficial fascia that lies superficial to the external oblique aponeurosis.

Scarpa's fascia attaches to what deep fascia in the leg?

Name the surgically important fibrous band extending from the xiphoid process to the symphysis pubis.

Through what triangular opening in the aponeurosis of the abdominal muscles do most herniations occur in males?

Card # 37

TIO

caput medusae

Scarpa's fascia

fascia lata

linea alba

superficial inguinal ring

Name the inferior free border of the aponeurosis of the external oblique.

The ilioinguinal nerve runs between what two abdominal muscles?

Name the serous membrane which lines the abdominal and pelvic cavities.

In what way do direct and indirect inguinal herniations differ anatomically?

What fibrous structure, enclosing the rectus abdominus, is formed by the three flat abdominal muscles?

Name the line formed by the inferior end of the posterior half of the rectus sheath.

Card # 38

- inguinal ligament
- internal oblique
  transversus abdominis
- peritoneum
- The indirect proceed via the inguinal canal whereas the direct proceed via the superficial ring only.
- rectus sheath
- arcuate line

Name the band of tissue that anchors the testis in the scrotal sac?

Name the structures that connect the head of the epididymis to the testis.

Name the numerous tubular structures found in the interior of the testis.

What term defines two layers of peritoneum that freely support the internal organs?

What do we call a large accumulation of serous fluid in the peritoneal cavity?

What fibrous structure divides the liver into right and left lobes?

- gubernaculum testis
- efferent ductules
- seminiferous tubules
- mesentery
- ascites
- falciform ligament

What organ, adjacent to the visceral surface of the liver, concentrates and stores bile?

What lymphatic organ lies in the upper left quadrant of the abdominal cavity?

The small intestines empties into what large intestinal structure?

What is another name for the GI tract?

How does the greater omentum work to prevent peritonitis?

What part of the small intestine is only partially attached to the mesentery?

Card # 40

- gall bladder
- spleen
- cecum
- alimentary canal
- seals off inflamed areas
- duodenum

Which sections of the small intestine are mobile due to the mesenteric attachment?

Name the mesentery of the appendix.

Refer to figure 3:1

What parts of the large intestine have a mesentery?

What parts of the large intestine lack a mesentery?

The cremasteric muscle arises from what muscles?

The deep inguinal ring lies between what two landmarks?

Card # 41

ileum
jejunum

---

mesoappendix

---

appendix
cecum
sigmoid colon
transverse colon

---

ascending colon
descending colon
rectum and anal canal

---

internal oblique
transversus abdominis

---

anterior superior iliac spine
pubic tubercle

What nerve travels through the spermatic cord?

What are the ridge-like structures found in the small intestine?

Chapter 4

What is the boundary between the greater and lesser pelvis?

Into which pelvis, greater or lesser, does the peritoneal cavity extend?

An imaginary line connecting the two ischial tuberosities divides the perineal region into what two triangular areas?

Name the bones forming the pelvis.

Card # 42

- genito femoral
- plicae circularis
- pelvic brim
- lesser pelvis
- anal triangle
  urogenital triangle
- right and left innominate bones
  sacrum and coccyx

Describe the position, relative to each other, of the anterior superior iliac spines and the upper aspects of the symphysis pubis in a person standing erect.

The obturator nerve and vessels transverse a canal travelling through what large foramen?

What ligament connects the coccyx and sacrum to the ischial spine?

What two foramina are formed by the sacrospinous and sacrotuberous ligaments?

How does the pubic arch differ between males and females?

Through what holes do the ventral and dorsal rami Sl to S4 exit the skeleton?

Card # 43

- all lie on the same vertical plane
- obturator foramen
- sacrospinous ligament
- greater and lesser sciatic foramina
- The pubic arch is wider in females.
- ventral and dorsal sacral foramina, respectively

What structure does the meeting of the ilium, ischium, and pubis form?

What purpose does the fat-filled retropubic space fulfill?

The obturator artery, after branching from the internal iliac artery, anastomoses with what branch of the external iliac artery?

The levator ani arises partially from the tendinous arch, a thickening of which muscle's fascia?

Name the three muscles that comprise the levator ani.

Name the u-shaped sling of fibers, originating from the pubococcygeus muscle, that relaxes during defecation.

Card # 44

- acetabulum
- accommodation of bladder expansion
- inferior epigastric a.
- obturator internus
- coccygeus
  iliococcygeus
  pubococcygeus
- puborectal sling

What fat-filled space accommodates the expanded rectum?.

Name the three parts of the sphincter ani externus.

The pudendal canal, transmitting the pudendal nerve and internal pudendal vessels, arises from what facial thickening?

Name the sacral origins of the pudendal nerve and of the pelvic parasympathetics (nervi erigentes).

What ligament anchors the penis to the symphysis pubis?

What muscle covers the crus of the penis?

Card # 45

- ischiorectal fossa
- subcutaneous
  superficial
  deep
- obturator internus fascia
- S2, 3; and 4
- suspensory ligament of the penis
- ischiocavernosis

Name the inferior fascial layer of the U.G. diaphragm.

On what structure does the prostate rest?

Through what structure(s) does sperm flow en route from the epididymis to the seminal vesicle?

The ejaculatory duct originates from the union of what two structures?

Refer to figure 4:1

Name the major branches of the internal iliac artery?

Card # 46

perineal membrane

pelvic diaphragm

ductus (vas) deferens
ampulla

ampulla
duct of the seminal vesicle

ilio-lumbar a.
internal pudendal a.
lateral sacral a.
inferior vesical a.

superior gluteal a.
superior vesical a.
inferior gluteal a. obturator a.
middle rectal a. (males)
uterine a. (females)

The piriformis muscle exits the pelvis via what foramen?

L4 and L5 contributions arrive to the sacral plexus via what nerve?

What are the origins of the sciatic nerve?

The dorsal arteries of the penis branch from what artery?

The dorsal nerves of the penis branch from what nerve?

What perineal structure provides the major support of the pelvic viscera?

Card # 47

- greater sciatic foramen
- lumbosacral trunk
- ventral rami L4 through S3
- internal pudendal
- pudendal
- perineal body

What muscle aids in emptying the spongy urethra?

Name the three parts of the male urethra?

Name the structure in the male corresponding to the uterus in the female?

What structures of the inferior wall of the bladder demarcate the trigone?

What structures unite the inferior aspect of the anal columns?

What vertebral levels contribute to the sacral plexus'?

- bulbospongiosus
- prostatic
  membranous
  spongy
- utricle
- two orifices of the ureters
  internal urethral orifice
- anal valves
- L4, 5 and Sl, 2, 3, 4

What are the anterior and posterior boundaries of the pelvic outlet of the female pelvis?

What organ is in contact with the anterior wall of the vagina? The posterior wall?

What structure is formed as the peritoneum passes from the uterus to the rectum?

What structure is formed as the peritoneum passes from the bladder to the uterus?

The uterine tube is contained within the free margin of what structure?

Name the peritoneal fold surrounding the uterine tube.

Card # 49

- lower-end of the symphysis
- tip of the coccyx

- bladder
- rectum and anal canal

- rectouterine pouch

- vesico-uterine pouch

- broad ligament

- mesosalpinx

Name the peritoneal fold that contains the ovary.

Name the most lateral part of the broad ligament that attaches to the pelvis.

What ligaments anchor the uterus in place?

Name the triangular-shaped thickening of the pelvic fascia lateral to the cervix and vagina.

Does the uterine artery pass above or below the ureter near the lateral fornix of the vagina?

What three structures transverse the female pelvic diaphragm?

Card # 50

mesovarium
suspensory ligament of the ovary
round sacro-uterine transverse cervical
transverse-cervical ligament (cardinal ligament)
above
urethra vagina rectum

What muscle is responsible for shrinking the vaginal orifice?

What structures in the female contain erectile tissue?

The uterus and vagina normally intersect at an angle of approximately 90 degrees. What is the name given to this orientation?

Name of the opening of the cervix into the vagina.

What are the divisions of the uterine tube from lateral to media?

What is the name of the paired gland that provides lubrication to the vaginal orifice during sexual stimulation?

Card # 51

- bulbospongiosus
- bulb of the vestibule (glans clitorus) crus of the clitorus (corpus cavernosum)
- anteversion
- external os
- fimbria
  infundibulum
  ampulla
  isthmus
  intramural
- greater vestibular gland

The round ligament of the uterus passes over what artery as it enters the deep inguinal ring?

What is the name of the region of the vagina around the cervical opening?

What two arteries freely anastomose along the lateral borders of the uterus?

What female erectile organ is homologous to the male penis?

The gubernaculum persists as what two structures in the female?

The levator ani and the voluntary external anal sphincter are innervated by which pudendal nerve branch?

Card # 52

- inferior epigastric
- fornix
- uterine and ovarian
- clitorus
- ligament of the ovary
  round ligament of the uterus
- inferior rectal n.

Is contraction of the sphincter urethrae voluntary or involuntary?

Which part of the urethra in the male is surrounded by the external urethral sphincter?

Which branch of the pudendal nerve innervates the sphincter urethrae muscle?

What type of nervous system outflow causes erection? Ejaculation?

What structures tend to oppose the rotation of the sacrum upward?

Name the caudal opening of the sacral canal?

Card # 53

voluntary
membranous
perineal n.
parasympathetic sympathetic
sacrospinous ligament sacrotuberous ligament
sacral hiatus

What ligament anchors the prostate to the pubic bone?

What muscle(s) stretch between the ischial rami?

Name the four divisions of the cervix.

Name the term describing the slight forward bending of the fundus and body of the uterus.

What areas of the alimentory canal are innervated by the pelvic splanchnic nerves.

Chapter 5

The round ligament of the uterus passes over what artery as it enters the deep inguinal ring?

Card # 54

puboprostatic ligament
sperficial and deep perineal muscles
internal os supravaginal cervix vaginal cervix external os
anteflexion
large intestine from the left colic flexure to the upper half of the anal canal
inferior epigastric

What is the longest vein in the body?

How many valves are found in the great saphenous vein?

Name the cutaneous nerves outlining the course of the sartorius muscle

The great saphenous vein pierces the fascia lata and empties into which vein?

Identify the clinical condition of abdominal visceral protrusion into the femoral canal.

What structures are contained in the cone-shaped femoral sheath?

Card # 55

- great saphenous
- 10 to 20
- lateral, intermediate, and medial cutaneous nerves of the thigh
- femoral vein
- femoral hernia
- femoral artery, vein and lymphatic vessels

What are the boundaries of the femoral triangle?

What are the boundaries of the femoral ring?

What nerve lies lateral to the femoral artery?

What are the three major branches of the femoral artery, arising either independently or from a common branch?

What fascial canal begins at the femoral triangle apex and ends at the hiatus of the adductor magnus?

What muscle, enclosed by the fascia lata, pulls on the iliotibial tract when contracted?

Card # 56

adductor longus m.
inguinal ligament
sartorius m.

---

ventral: inguinal ligament
dorsal: pectineus muscle and fascia
lateral: femoral vein
medial: lacunar ligament

---

femoral

---

lateral and medial femoral circumflex a.
profunda femoris a.

Hunter's canal

---

tensor fascia lata

As the femoral artery passes through the adductor hiatus, it changes its name to what artery?

Describe the origin of the rectus femoris muscle.

The adductor muscle group inserts onto what common ridge?

The profunda femoris artery travels between what two adductor muscles?

The tendons of what four muscles from the ligamentum patellae?

The femoral nerve innervates what lower limb muscles?

Card # 57

popliteal	
straight head: anterior inferior iliac spine reflected head: acetabulum	
linea aspera	
adductor longus pectineus	
quadriceps femoris: rectus femoris vastus lateralis vastus intermedius vastus medialis	
iliacus pectineus quadriceps femoris sartorius	

Name the muscles innervated by the obturator nerve?

On what tubercle of the medial epicondyle of the femur is the insertion for the hamstring component of adductor magnus muscle?

What sesamoid bone forms in the quadriceps femoris tendon?

What muscle receives innervation from both the obturator and sciatic nerves?

Refer to figure 5:1

What structure(s) traverse the greater sciatic foramen above the piriformis?

What structure(s) traverse the greater sciatic foramen below the piriformis muscle?

Card # 58

- adductor group
- gracilis m.
- obturator externus m.

- adductor tubercle

- patella

- adductor magnus

- superior gluteal nerve and vessels

- inferior gluteal nerves and vessels
- posterior cutaneous nerve of the thigh
- pudendal nerve and internal pudendal vessels
- sciatic nerve

Which muscle(s) are innervated by the inferior gluteal nerve?

Which muscle(s) are innervated by the superior gluteal nerve?

What is the largest nerve in the body?

What structure traverse the lesser sciatic foramen?

What muscles insert into the iliotibial tract?

What structure allows the gluteus maximus to slide over the greater trochanter of the femur?

Card # 59

gluteus maximus
gluteus medius and minimus tensor facia lata
sciatic (L4, 5; Sl, 2, 3)
internal pudendal vessels obturator internus m. pudendal n.
gluteus maximus tensor fascia lata
trochanteric bursa

What common action is performed by the muscles innervated by the superior gluteal nerve?

The obturator internus tendon lies between what two other lateral rotators of the femur?

Where in the gluteal region should an intramuscular injection be administered?

What muscles form the borders of the diamond-shaped popliteal fossa?

What two nerves contribute to the formation of the sural nerve, the nerve that takes over for the posterior cutaneous nerve of the thigh?

The tendo calcaneus forms from what two muscles?

Card # 60

- abduction of the femur
- superior and inferior gemelli m.
- upper outer quadrant
- biceps femoris
  gastrocnemius (2 heads)
  semimembranosus
  semitendinosus
- medial and lateral sural cutaneous nerves
- gastrocnemius
  soleus

The sciatic nerve gives rise to what two major nerves in the posterior thigh?

What are the two main actions of the hamstring muscles?

Name the muscles that comprise the hamstrings.

From what common site do the hamstrings originate?

Why doesn't the short head of biceps femoris belong to the hamstring group?

What nerve innervates the hamstring?

Card # 61

common peroneal tibial	
flexion of the knee joint extension of the hip joint	
adductor magnus biceps femoris (long head) semimembranosus semitendinosus	
ischial tuberosity	
originates from the shaft of the femur innervated by the common peroneal n.	
sciatic	

Which superficial group of muscles in the posterior compartment of the leg receives innervation from the tibial nerve?

What leg muscles receive innervation from the deep peroneal nerve?

What are the two main terminal branches of the common peroneal nerve?

What vessel is the major supplier of blood to the hamstrings?

The popliteal artery is a continuation of what artery?

The muscles of the anterior compartment of the leg mainly contribute to what action?

Card # 62

- gastrocnemius
- plantaris
- soleus

---

anterior- leg compartment:
- extensor digitorum longus
- (also peroneus tertius)
- tibialis anterior
- extensor hallucis longus

---

superilicial arid deep peroneal nerves

---

profunda femoris artery
(perforating branches)

---

femoral

---

dorsiflexion (extension)

What are the main terminal branches of the popliteal artery?

What two muscles, both innervated by the superficial peroneal nerve, arise from the fibula?

What branch of the posterior tibial artery feeds the lateral compartment of the leg?

What are the major actions of peroneus longus and brevis?

Name the thickening of deep fascia, shaped like a "Y", that holds tendons on the dorsum of the foot in place.

The dorsalis pedis artery is a continuation of what artery?

Card # 63

- anterior and posterior tibial arteries
- peroneus longus and brevis
- peroneal a.
- eversion and plantar flexion of the foot
- inferior extensor retinaculum
- anterior tibial

What muscle, actually part of the extensor digitorum longus muscle inserts into the fifth metatarsal bone?

The plantar arterial arch is completed by a deep plantar branch of what artery?

Which flexor of the foot grooves the tibial malleolus?

Which flexor uses the sustentaculum tali as a pulley?

Which deep group muscles of the posterior compartment of the leg receive innervation from the tibial nerve?

What nerve innervates extensor digitorum brevis and hallucis brevis?

Card # 64

- peroneus tertius
- dorsalis pedis
- tibialis posterior m.
- flexor hallucis longus m.
- flexor digitorum longus
  flexor hallucis longus
  popliteus
  tibialis posterior
- deep peroneal

Damage to what nerves can lead to "foot drop"?

Which two long flexor muscles have tendons that, due to the locations of their insertions, cross on the sole of the foot?

What are the weight bearing points of the foot?

What ligaments help to maintain the longitudinal arch of the foot?

The tibial nerve branches into two nerves in the sole of the foot?

Name the origin and insertion of the quadratus plantae.

Card # 65

- common or deep peroneal nerves

- flexors digitorum longus and hallucis longus

- calcaneus
  heads of metatarsal bones one and five

- plantar aponeurosis
  spring ligament (calcaneonavicular)

- medial and lateral plantar

- o: calcaneus
  i: flexor digitorum longus m.

What is the function of the four dorsal interossei muscles? The three plantar interossei?

Which muscles in the sole of the foot are innervated by the medial plantar nerve?

Name the ligaments that support the knee joint medially and laterally.

Which ligament connects the sustentaculum tali to the navicular bone?

Which knee joint ligament prohibits extreme knee extension?

Why is the medial meniscus more prone to injury than the lateral meniscus?

Card # 66

abduction of the toes

adduction of the toes

---

abductor hallucis
first lumbrical
flexor digitorum brevis
flexor hallucis brevis

---

medial: tibial collateral
lateral: fibular collateral

---

spring ligament

---

anterior cruciate ligament

---

the medial meniscus is attached to the tibial collateral ligament
the lateral meniscus is freely mobile

What is the strongest ligament of the hip joint that contributes to the formation of the joint capsule?

What muscles are innervated by the nerve to the quadratus femoris?

What muscles are innervated by the nerve to the obturator internus?

Name the major superficial cutaneous nerves of the buttock and posterior thigh?

The largest branch of the femoral nerve does not enter the adductor hiatus. Name this branch.

Lateral thigh rotators tend to insert onto which port of the femur?

Card # 67

- iliofemoral ligament

- inferior gemellus
  quadratus femoris

- obturator internus
  superior gemellus

- middle and superior cluneal n.
  posterior cutaneous nerve
  of the thigh

- saphenous nerve

- greater trochanter

Name the three tendons of the pes anserinus.

What additional action is provided for by the rectus femoris muscle other than extension of the leg in conjunction with the rest of the quadriceps?

The medial 2/3 of the sole of the foot receives innervation from what nerve? Lateral 1/3 of the sole?

The oblique popliteal ligament, an extracapsular knee joint structure, is derived from which muscle?

What two arteries feed the head of the femur?

Which artery supplies the cruciate ligaments?

Card # 68

- gracilis
- sartorius
- semitendinosis

---

flexion of the hip joint

- medial plantar
- lateral plantar

---

semimembranosus

---

medial and lateral femoral circumflex

---

middle genicular

Through what fibrous structure does the anterior tibial artery pass on its way to the extensor compartment of the leg?

Which group of back muscles anchor the upper extremities to the axial skeleton?

Chapter 6

What is the major function of the intermediate group of back muscles?

Which nerve rami, ventral or dorsal, innervate the superficial group of back muscles? Intermediate group? Deep group?

Why must a lumbar puncture be performed caudal to the conus medullaris?

Into which lumbar interspace is a needle usually introduced for retrieval of CSF?

Card # 69

- interosseous membrane
- superficial
- respiration
- ventral
  ventral
  dorsal
- to prevent damage to the spinal cord (CNS)
- fourth

What is the term for the ventral and dorsal roots of the spinal cord inferior to the conus medullaris?

What muscles form the borders of the triangle auscultation, an area of clinical diagnostic importance?

What are the names of the dorsal rami of C1 and C2?

The levator scapulae muscle, originating as four tendons from the upper four cervical transverse processes, is innervated by what nerve(s)?

The subarachnoid space of the spinal cord contains what fluid?

Name the muscles comprising the deep group of the back.

Card # 70

- cauda equina
- latissimus dorsi
  rhomboid major
  trapezius
- suboccipital nerve
  greater occipital nerve
- cervial plexus (C3, 4)
  dorsal scapular (C5)
- cerebrospinal fluid (CSF)
- erector spinae
  semispinalis capitis
  splenius capitis

What membrane, continuous from the cranium, invest the spinal cord?

The cervical enlargement of the spinal cord corresponds to the origin of which nervous plexus?

Name the prolongations of the pia mater that anchor the spinal cord.

How many pairs of spinal nerves are there generally?

At what level does the intradural filum terminate end?

What three muscles contribute to the suboccipital triangle?

Card # 71

pia mater
brachial plexus
denticulate ligaments
31 pairs (8 cervial, 12 thoracic, 5 lumbar, 5 sacral, and 1 coccygeal)
S2
obliquus capitis inferior obliquus capitis superior rectus capitis posterior major

Name the gelatinous substance found within the annulus fibrosus.

What longitudinal ligament extends from the back of the body of the axis to the triangular sacral canal?

Which type of joint exists between vertebral neural arches?

What yellow ligaments, connecting the front of the upper lamina to the back of the lower lamina, provide antigravity support?

What ligaments attach the tips of adjacent spinous processes of the vertebrae?

What veins drain the red bone marrow rich vertebral bodies?

Card # 72

- nucleus pulposus
- posterior longitudinal ligament
- gliding synovial joint
- ligainenta flava
- supraspinous ligaments
- basivertebral

Which vertebra has the most prominent spinous process (vertebra prominens)

What ligament covers the spines of vertebrae Cl through C6?

How many cervical vertebrae are there in the vertebral column? Thoracic? Lumbar? Sacral?

Chapter 7

The trigeminal nerve supplies cutaneous innervation to the face. Name its three divisions.

Describe the anatomical lines separating the dermatomes of the three divisions of the trigeminal nerve.

Name the facial bones bordering the maxilla.

Card # 73

C7

ligamentum nuchae

7
12
5
5

ophthalmic (VI)
maxillary (V2)
mandibular (V3)

VI from V2: from just below the tip of the nose to the medial angle of the eye
V2 from V3: from the corner of the mouth to a point midway between the eye and ear

frontal
lacrimal
nasal
zygomatic

The external auditory meatus is part of what bone?

What two processes form the zygomatic arch?

What two bones are separated by the coronal suture?

Name the major cutaneous branches of V1.

Name the major cutaneous branches of V2.

Name the major cutaneous branches of V3.

Card # 74

temporal

zygomatic process of the
temporal bone
temporal process of the
zygomatic bone

frontal and parietal

external nasal n.
infra-trochlear n.
lacrimal n.
supra-orbital n.
supra-trochlear n.

infra-orbital n.
zygomatico-facial n.
zygomatico-temporal n.

auriculo-temporal n.
buccal n.
mental n.

Describe the course of the parotid duct.

What artery and nerve run just superior to the parotid duct?

Within the parotid gland, the facial nerve divides into numerous branches to supply the facial muscles of expression. Name these major branches.

The buccal branch of the trigeminal nerve pierces the buccinator muscle to innervate what area?

Where does the facial artery cross the border of the mandible?

What cartilages form the nostrils?

Card # 75

Runs lateral to the masseter m., turns inward at the anterior border of the masseter m., pierces the buccinator m., and then opens opposite the upper 2nd molar

transverse facial a.
zygomatic branch of
the facial n.

temporal
zygomatic
buccal
mandibular
cervical

buccal mucosa of the mouth

the anterior border of the masseter m.

alar cartilage

What structure forms the anterior portion of the nasal septum?

Name the potential space between the eyeball and eyelids. What membrane lines this space?

Name the red triangular prominence near the medial canthus of the eye.

The lacrimal canaliculus opens near the caruncula and drains lacrimal fluid to what structure?

Name the transparent outer coating of the anterior aspect of the eyeball.

What glands lie posterior to the eyelids?

Card # 76

- septal cartilage
- conjunctival sac conjunctiva
- caruncula lacrimalis
- lacrimal sac
- cornea
- tarsal (meibomian)

Name the white posterior portion of the outer coat of the eyeball.

What fibro band acts as part of the orbicularis oculi muscle?

What is an inflammation of a small sebaceous gland around the follicle of an eyelash known as?

The lacrimal sac drains through the nasolacrimal duct. Where does this duct then enter the nose?

The lacrimal gland lies in what part of the orbit?

What two muscles are united by the broad epicranial aponeurosis?

Card # 77

- sclera
- medial palpebral ligament
- a stye
- inferior meatus
- upper and outer
- frontalis m. occipitalis m.

What lies in the clinically important area between the pericranium and the epicranial aponeurosis?

What two bones fuse at the root of the nose to form the nasion?

What important bony protuberance serves as the insertion for the occipitalis and trapezius muscles?

What layer of the scalp contains the blood vessels and cutaneous nerves?

Suture lines separate the individual bones of the skull. Name the point where the sagittal and lambdoidal sutures meet. The sagittal and coronal sutures.

Name the intermediate spongy bone layer of the calvaria.

Card # 78

- loose areolar tissue (infection often spreads in this layer)
- nasal
- external occipital protuberance (inion)
- connective tissue
- lambda
  bregma
- diploe

Through what foramina in the posterior fossa do cranial nerves VII and VIII exit? Nerves IX, X and XI?

The middle meningeal as well as the recurrent branch of the mandibular nerve traverse what cranial foramen?

Cranial nerve XII exits the skull through what structure?

On the interior surface of the skull, what bony landmark lies at the grooves of the superior sagittal and transverse sinuses?

Name the three membranous layers forming the meninges.

What dural layer splits to enclose venous sinuses?

Card # 79

internal acoustic meatus
jugular foramen

foramen spinosum

hypoglossal canal

internal occipital protuberance

dura mater
arachnoid membrane
pia mater

dura mater
(inner and outer layers)

Name the clinically serious, even life-threatening accumulation of blood between the skull and dura. What type of blood accumulates?

What artery supplies the dura mater and many of the cranial bones?

Name the numerous bush-like structures that drain CSF into venous sinuses.

Name the semicircular fold of the inner dural layer that separates the two cranial hemispheres.

Name the dural fold that separates the cerebellum from the occipital poles of the cerebral hemispheres.

Name the clinically serious condition of bleeding into the potential space between the dura mater and arachnoid membrane. What type of blood accumulates?

Card # 80

- epidural hematoma arterial
- middle meningeal
- arachnoid granulations
- falx cerebri
- tentorium cerebelli
- subdural hematoma venous

What CSF containing space lining the brain acts as a shock absorber?

A substantial space between pia and arachnoid membranes is known as a cistern. What is the largest such cistern?

What meningeal layer closely follows the surface of the brain?

The inferior sagittal sinus and the great cerebral vein join to form what structure?

What bony structure transmits the fibers of the olfactory nerve through its numerous tiny foramina?

Name the triangular projection of the ethmoid bone that acts as the anterior attachment of the falx cerebra.

Card # 81

- subarachnoid space
- cisterna magna (cisterna cerebello-medullaris)
- pia mater
- straight sinus
- cribriform plate of the ethmoid bone
- crista galli

The right and left vertebral arteries join to form what vessel?

What four large arteries join to form the circle of Willis?

The basilar artery splits to form what major vessels?

Name the two unpaired arteries in the circle of Willis.

Name the five terminal branches of the internal carotid artery.

The anterior ethmoidal foramen transmits what structures?

Card # 82

basilar a.
two vertebral
two internal carotid
anterior inferior cerebellar a.
superior cerebellar a.
posterior cerebral a.
pontine a.
labyrinth a.
anterior communicating
basilar
anterior cerebral a.
middle cerebral a.
posterior communicating a.
ophthalmic a.
anterior choroid a.
anterior ethmoidal nerve and artery

The superior orbital fissure transmits what structures?

The optic nerve and the ophthalmic artery enter the orbit through what opening?

The maxillary nerve traverses what cranial foramen?

The foramen ovale transmits what important nerve and vessel?

What structures run across the foramen lacerum?

What two structures join to form the jugular vein? Where is it formed?

Card # 83

cranial nerves III, IV, VI, and VI superior ophthalmic v. sympathetic fibers orbital branch of the middle meningeal a.	
optic canal	
foramen rotundum	
mandibular n. (V3) accessory meningeal a.	
nerve of the pterygoid canal internal carotid a.	
inferior petrosal sinus sigmoid sinus jugular foramen	

What major structures transverse the foramen magnum?

What three bones form the anterior cranial fossa

What bony landmark forms the most anterior border of the posterior fossa?

Name the inclining bony surface anterior to the foramen magnum.

Blood drains from the cavernous sinus via what three structures?

The pituitary gland rests in what bony recess?

Card # 84

accessory n. (XI) anterior and posterior spinal arteries medulla vertebral arteries
ethmoid frontal sphenoid
dorsuam sella
clivus
a small vein destined for the pterygoid plexus superior and inferior petrosal sinuses
hypophyseal fossa (sella turcica)

What is the name of the circular dural fold which covers the hypophyseal fossa?

Which cranial sinus drains blood from the transverse sinus?

What structures traverse the cavernous sinus?

What two nerves form the nerve of the pterygoid canal?

The pterygoid canal connects what two foramina?

Name the gap in the orbit between the maxilla and the greater wing of the sphenoid bone.

Card # 85

- diaphragma sellae
- superior petrosal sinus
- internal carotid a.
  cranial nerves III, IV, V_1, V_2 and VI.
  sympathetic plexus
- greater petrosal (parasympathetic)
  deep petrosal (sympathetic)
- foramen lacerum
  sphenopalatine foramen
- inferior orbital fissure

Refer to figure 7:1

What bones contribute to the formation of the orbital cavity?

Which wall of the orbit is paper thin?

Name the tough ring serving as the origin of the four orbital recti muscles.

The ophthalmic nerve (VI) gives off what branches within the orbit?

Which orbital muscles receive innervation from the oculomotor nerve

Which orbital muscle receives innervation from the trochlear nerve (IV)?

Card # 86

	ethmoid, lacrimal, sphenoid, frontal, maxilla, zygomatic
	medial wall (lamina papyracea)
	anulus tendineus
	frontal n. lacrimal n. nasociliary n.
	inferior oblique levator palpebra superious superior, middle, and inferior recti
	superior oblique

Which orbital muscle receives innervation from the abducens nerve (VI)?

What vein drains the upper portion of the face to the cavernous sinus?

The optic nerve contains an artery in its center. Name this artery.

Name the four refractive layers in the eye through which light passes before striking the retina.

What is the name of the retinal region where the optic nerve enters the eyeball?

In the fundus of the eye, the blood vessels emerge from and drain toward what structure?

Card # 87

lateral rectus

superior ophthalmic 1.

(important clinically as a route of infection)

central artery of the retina

cornea
aqueous humor
lens
vitreous humor

optic disc (blind spot)

optic papilla

The eyeball has a tough outer layer (sclera), a middle vascular layer, and an inner neural layer (retina). Name the middle vascular layer.

Name the area of the retina responsible for the highest visual acuity. Only cones lie in this area.

What neck muscles have two bellies separated by an intertendon?

Five large nerves emerge in the neck along the posterior border of the sternocleidomastoid muscle. Name them.

The external jugular vein crosses the sternocleidomastoid muscle and then dips behind the clavicle to join what vein?

The anterior and middle scalene muscles form the interscalena triangle with rib 1. What two important structures pass through this triangle?

Card # 88

choroid

fovea centralis

digastric omohyoid

accessory (XI) great auricular lesser occipital supraclavicular transverse cervical

subclavian

brachial plexus subclavian a.

What nerve descends vertically across the surface of the anterior scalene on its way to the thorax?

Irritation of the phrenic nerve may refer pain to what cutaneous area?

What are the boundaries of the posterior triangle of the neck?

What are the boundaries of the anterior triangle of the neck?

What structures are held together by the carotid sheath?

What bony structures do the vertebral arteries ascend in the neck?

Card # 89

- phrenic (C3, 4, 5)
- clavicular and shoulder (supraclavicular n. is C3, 4; phrenic n. is C3, 4, 5)
- middle 1/3 of clavicle
  sternocleidomastoid m.
  trapezius m.
- mandible (inferior border)
  median line of the neck
  sternocleidomastoid (anterior border)
- common carotid a.
  internal jugular v.
  vagus n.
- foramina transversaria

What cartilage forms the laryngeal prominence (Adam's apple)?

What bones lies at the angle between the upper neck and the floor of the mouth?

What cartilagenous structure lies inferior to the thyroid cartilage?

The sternocleidomastoid muscle splits into two heads. On what bone does each insert?

What are the boundaries of the muscular triangle of the neck?

What are the boundaries of the submandibular triangle?

Card # 90

- thyroid cartilage
- hyoid bone
- cricoid cartilage
- clavicle
  manubrium
- midline of the neck
  sternocleidomastoid m.
  superior belly of the omohyoid m.
- anterior and posterior bellies
  of the digastric muscle
  inferior border of the mandible

What are the boundaries of the carotid triangle?

Name the four thin straplike muscles that run vertically in the anterior triangle of the neck.

The strap muscles receive their innervation from levels Cl thorugh C3 via what neural structure?

The trachea lies just beneath the skin between what glandular and bony structures?

What nerve pierces the thyrohyoid membrane? What area does it innervate?

What nerve innervates the cricothyroid muscle?

Card # 91

- posterior belly of the digastric m.
- sternocleidomastoid m.
- superior belly of the omohyoid m.

- sternohyoid
- sternothyroid
- superior belly of the omohyoid
- thyrohyoid

---

ansa cervicalis

---

beneath the thyroid gland and above the manubrium of the sternum.

---

internal laryngeal

upper portion of the larynx and pyriform fossa

---

external laryngeal

The common and internal carotid arteries give off no branches in the neck. The external carotid artery gives off six. Name 3 of them.

Name three others.

Name the blood pressure sensitive organ within the bifurcation of tile common carotid artery? The chemoreceptive organ?

What cranial nerve innervates the carotid body and sinus?

Name the long, sharp protuberance on the inferior aspect of the temporal bone.

What gland wraps around the free posterior border of the mylohyoid muscles?

Card # 92

- ascending pharyngeal a.
- facial a.
- lingual a.

- occipital a.
- superior thyroid a.
- posterior auricular a.

- carotid sinus
- carotid body

- glossopharyngeal (IX)

- styloid process

- submandibular gland

The intermediate tendon of the digastric pierces what small muscle?

What two nerves innervate the digastric muscle?

What nerve ascends alongside the trachea just posterior to the thyroid gland? What is it responsible for?

Name the small endocrine glands that lie posterior to the thyroid gland between its capsule and sheath.

The carotid tubercle lies on what vertebra?

The laryngeal prominence lies at what vertebral level?

Card # 93

	stylohyoid
	anterior belly: V3 posterior belly: VII
	recurrent laryngeal motor innervation to the laryngeal muscles
	parathyroid glands
	C6
	C6

Characteristically the facial muscles of expression do not arise and insert on bony structures. They typically remain within what tissue layer?

What are the boundaries of the suprahyoid triangle?

What muscles form the floor of the posterior triangle of the neck?

What is the name of the large, sheet-like muscle extending from the mandible to below the clavicle? What is its innervation?

What two veins form the external jugular vein?

The spinal accessory nerve (XI) innervates what two muscles?

Card # 94

superficial fascia
anterior belly of the digastric m. hyoid bone midline of the neck
anterior, middle and posterior scalene levator scapulae splenius capitis
platysma facial n. (VII)
retromandibular posterior auricular
sternocleidomastoid trapezius

Name the branches of the subclavian artery?

In the neck, the sympathetic trunk courses posterior to what large cranial nerve?

What muscles extend from the interior surface of the mandible to the hyoid bone?

The external investing fascia of the neck splits to enclose what two muscles?

The prevertebral fascia encloses what structures?

What major vein descends in the neck along the ventral surface of the sternocleidomastoid muscle?

Card # 95

costocervical trunk internal thoracic a. thyrocervical trunk vertebral a.
vagus (X)
geniohyoid mylohyoid
sternocleidomastoid trapezius
vertebral column and its muscles
external jugular

What two veins unite to form the brachiocephalic vein?

Into what vessel does the thoracic duct empty?

Name the insertion of the longus capitis muscle.

What bone forms the inferior portion of the nasal septum?

What important nerve, vein and artery pass through the tissue of the parotid gland?

Name the insertion of the temporalis muscle.

Card # 96

internal jugular subclavian
left brachiocephalic v. (in the angle between the left subclavian v. and left internal jugular v.)
C3, 4, and 5 carotid tubercle (C6)
vomer
facial n. retromandibular v. external carotid a.
coronoid process of the mandible

The mandibular notch lies between the head and the coronoid process of the mandible. What nerve and vessels pass through this notch?

What ligament attaches to the lingula of the mandible?

The mandibular foramen transmits what nerve and vessels?

What structures form the walls of the infratemporal fossa?

The mouth opens in a hinge-like motion through what joint?

The articular disc adds versatility to the temporomandibular joint. What three types of movements can this joint accomplish?

Card # 97

- masseteric n. and vessels

- sphenomandibular ligament

- inferior alveolar n. and vessels

- anterior: maxilla
  lateral: mandibular ramus
  medial: medial pterygoid plate
  superior: greater wing of the sphenoid bone

- temporomandibular joint

- hinge (open and close)
  move from side to side
  protraction and retraction

What nerve(s) branch to innervate the teeth of the mandible? The maxilla?

Name the small delicate nerve that joins the lingual nerve high in the intratemporal fossa.

What are the superior and inferior extents of the retropharyngeal space?

What fascial layers form the walls of the retropharyngeal space?

What is the name given to the vertebra CI? C2?

Name the bony projection of C2 which lies in the anterior arch of the atlas.

Card # 98

inferior alveolar
anterior, middle, and
posterior superior
alveolar

chorda tympani

superior: base of the skull
inferior: diaphragm

buccopharyngeal
prevertebral

atlas

axis

odontoid process (dens)

Name the strong, curved ligament that tightly holds the dens of the axis to the anterior arch of the atlas.

Name the joint between the occipital condyle and the superior articular facet of the atlas.

The transverse ligament and the upper and lower bands are known collectively as what ligament?

Name the upward continuation over the atlanto-occipital joint of the posterior longitudinal ligament of the vertebral column.

What ligament limits the side-to-side movement of the head?

What two structures are joined by the alar ligament?

Card # 99

- transverse ligament
- atlanto-occipital joint
- cruciform ligament
- membrana tectoria
- alar ligament
- dens
  lateral margin of the foramen magnum

What organ receives motor innervation from the hypoglossal nerve (XII)?

The glossopharyngeal nerve (IX) runs closely applied to what small muscle passing the internal and external carotid arteries?

There are gaps between the three pharyngeal constrictor muscles through which important nerves and vessels pass. What important structures pass between the inferior and middle constrictors?

What important nerve and muscle pass between the middle and superior constrictors?

What are the superior and inferior extents of the pharnyx?

What structure separates the nasopharynx from the oropharynx?

Card # 100

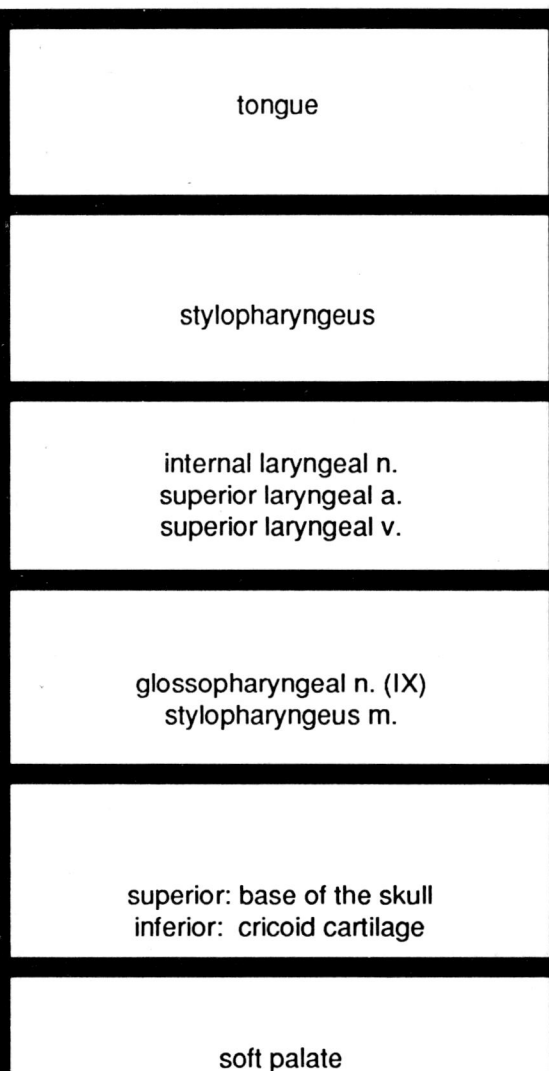

What structure separates the oropharynx from the larynx?

Name the lymphoid organ that lies between the palatoglossal and palatopharyngeal arches.

Name the boundary between the oral and pharyngeal parts of the tongue.

Name the various types of papillae on the tongue. Which type contains a large number of taste buds?

What are the three structural components of the nasal septum?

Name the curved slit found in the middle meatus of nose. Which ostia lies in it?

Card # 101

epiglottis

palatine tonsil

sulcus terminalis

circumvallate filiform fungiform circumvallate

perpendicular plate of the ethmoid bone septal cartilage vomer

hiatus semilunaris ostia for the maxillary, frontal, and ethmoidal sinus

Extraction of a molar or premolar tooth may rupture the wall of what sinus?

Name the four sets of sinuses.

What structures run through the greater palatine canal?

The auditory tube enters the nasopharynx cavity at the level of what nasal concha?

The greater palatine foramen lies medial to which tooth?

What two muscles meet at the pterygomandibular raphe?

Card # 102

| maxillary |

| ethmoidal
frontal
maxillary
sphenoidal |

| greater palatine n. and a. |

| inferior |

| 3rd molar |

| superior pharyngeal constrictor
buccinator |

What structures pass through the gap between the upper free border of the superior constrictor and the base of the skull?

What muscle elevates and retracts the soft palate during swallowing?

What bony structure acts as a pulley for the tensor palate muscle?

Name the median fold of mucous membrane that connects the tongue to the floor of the mouth.

What nerve spirals around the submandibular duct?

What muscle protrudes the tongue?

Card # 103

| ascending pharyngeal a. |
| auditory tube |
| levator palati m. |

| levator palate |

| hamulus of the medial pterygoid plate |

| frenulum linguae |

| lingual |

| genioglossus |

Name the anterior and then the posterior attachments of the vocal ligaments.

What muscle increases the length of the fibers and usually tenses the vocal cords?

What is the only muscle capable of opening the rima glottidis and thereby maintaining the respiratory airway?

The tympanic cavity is an air space within what skull bone?

The aditus connects what two air spaces?

Name the three auditory ossicles.

Card # 104

thyroid cartilage
arytenoid cartilages

cricothyroid

posterior cricoarytenoid m.

temporal

mastoid air cells
tympanic cavity

malleus
incus
stapes

The stapes vibrates what transducing membrane?

On what bone does the tensor tympani muscle insert?

What are the attachments of the pterygospinous ligament?

Which pterygoid muscle, lateral or medial, assists in opening and protracting the mouth?

Describe the lymphatic drainage from the tongue and lower oral cavity.

What nerve provides sensory innervation to the anterior 2/3 of the tongue?

Card # 105

oval window
malleus
posterior margin of the lateral pterygoid plate sphenoidal spine
lateral
inferiorly to the submandibular and deep cervical nodes. (commonly causes "swollen glands")
lingual

What nerve provides sensory innervation to the oropharynx?

The constrictor muscles of the pharynx are innervated by which cranial nerve?

Name the four layers of the pharyngeal wall (superficial to deep).

What cranial nerve innervates the tensor palatine muscle?

Refer to figure 7:2
The larynx consists of the three paired and three unpaired sets of cartilages. Name them.

What nerve provides sensory innervation to the larynx above the vocal cords? Below the vocal cords?

Card # 106

glossopharyngeal (IX)

vagus (X)

buccopharyngeal fascia
muscular layer
fibrous pharyngobasilar fascia
mucous layer

trigeminal (V3)

paired: arytenoid   unpaired: epiglottis
cuneiform   cricoid
corniculate   thyroid

internal laryngeal

recurrent laryngeal

The rotation of what laryngeal cartilage results in the medial and lateral displacements of the vocal cords?

What ligament forms the false vocal cords?

Name the elastic membrane connecting the vocal cords to the cricoid, thyroid, and arytenoid cartilages.

Chapter 8

Name the two major divisions of the autonomic nervous system.

The cell bodies of the preganglionic sympathetic neurons lie in the lateral horns of which spinal segments?

The axons of the preganglionic sympathetic neurons pass out of the ventral roots of which spinal segments?

Card # 107

- arytenoid
- vestibular ligament
- conus elasticus
- sympathetic parasympathetic
- C8 through L2
- Tl through L2

What type of nerve is formed by the union of the dorsal and ventral spinal roots?

The myelinated preganglionic sympathetic neuron passes through what neural structure immediately prior to its synapse in the sympathetic trunk?

How many sympathetic ganglia exist in the sympathetic trunk?

Which sympathetic ganglia receive white rami communicantes?

Every spinal nerve receives what type of communicating fiber from the sympathetic trunk?

The cell bodies of the unmyelinated postganglionic sympathetic neurons lie in what structure?

Card # 108

- spinal
- white ramus communicantes
- 3 cervical
  11 thoracic
  4 lumbar
  5 sacral
- T1 through L2
- gray ramus commuicans
- sympathetic ganglia

The eight cervical ganglia fuse into how many actual ganglionic structures? Name them.

Name the six different autonomic nervous plexus.

The lower thoracic sympathetic ganglia send three splanchnic nerves to which visceral nervous plexus?

Of the six sympathetic nervous plexuses, three receive only postganglionic sympathetic fibers, one receives only preganglionic, and two receive mixed. Name the plexuses receiving only postganglionic fibers.

The fibers coursing from the sympathetic trunk through the splanchnic nerves are (preganglionic/postganglionic/mixed)?

Sympathetic fibers destined for the superior and inferior hypogastric plexuses are (preganglionic/postganglionic/mixed).

Card # 109

three superior, middle, and inferior cervical ganglia	
coeliac cardiac pulmonary	esophageal superior hypogastric inferior hypogastric
coeliac	
cardiac esophageal pulmonary	
preganglionic	
mixed	

In which brainstem nuclei lie preganglionic parasympathetic cell bodies?

The cell bodies of preganglionic parasympathetic neurons lie at which spinal levels?

The sacral parasympathetic nervous contribution to the inferior hypogastric plexus arrives via what parasympathetic white rami?

Which division(s) of the autonomic nervous system distribute to the body wall? To the viscera?

Name the four cranial parasympathetic ganglia.

What classes of neurons pass through the cranial parasympathetic ganglia without synapsing?

Card # 110

Edinger-Westphal superior and inferior salivatory
S2, 3, and 4
nervi erigentes
sympathetic sympathetic and parasympathetic
ciliary otic pterygo-palatine submandibular
sensory sympathetic

What type(s) of fibers synapse in cranial parasympatetic ganglia?

The parasympathetic outflow via the ciliary ganglion supplies what area(s)? Pterygopalatine ganglion?

The parasympathetic outflow via the otic ganglion supplies what area(s)? Submandibular ganglion?

What nerve carries parasympathetic secretomotor fibers to the submandibular ganglion and taste fibers from the anterior-two-thirds of the tongue?

The cell bodies of the secretomotor neurons of the nervus intermedius lie in which brainstem nucleus?

Chapter 9

What are the four classifications of joints?

Card # 111

only parasympathetic

eyeball

glands of the palate, lacrimal gland, and glands of paranasal sinuses, nasopharynx, and nose

parotic, buccal, labial, and molar

glands in the floor of the mouth

chorda tympani n.

superior salivatory

primary cartilaginous
secondary cartilaginous
fibrous
synovial

To which joint class do the skull sutures belong?

The epiphysis of a long bone, where bone and hyaline cartilage meet, belongs to which class of joints?

An intervertebral disc and the symphysis pubis are both examples of which classification of joints?

Which class of joints is, in general, the most mobile?

Name the six types of synovial joints.

List the possible movements of a condylar joint.

- fibrous
- primary cartilaginous
- secondary cartilaginous
- synovial
- hinge, saddle, ball and socket, pivot, condylar, gliding
- flexion, extension, abduction, adduction, circumduction

An interosseous membrane is an example of what type of fibrous joint?

In what class of joints do you find fibrocartilage sandwiched between two bony surfaces covered with hyaline cartilage?

The surface of bones or cartilages are joined only by fibrous tissue in what class of joints?

What tissue structures occupy space in joints formed by the lack of bone articulation?

The elbow joint is a hinge-type synovial joint since it allows which types of movements?

Which specific joint is the most mobile?

Card # 113

- syndesmosis
- secondary cartilaginous
- fibrous
- fatty pads
  fibrocartilaginous discs
- flexion and extension
- shoulder joint

The ball and socket hip joint is less mobile than the shoulder joint due to its increase in which joint function?

The joint at the base of the thumb can move in two planes. What type of synovial joint is it?

Chapter 10

The urachus is the vestige of what primitive structure?

What ligament is the vestige of the urachus?

The falciform ligament of the liver contains what fibrous cord?

The ligamentum teres is the remnant of what fetal vein?

Card # 114

- stability
- saddle
- allantois
- median umbilical ligament
- ligamentum teres
- left umbilical v.

The ductus venosus of the fetus becomes what vestigal structure in the adult?

The umbilical arteries are branches of what major artery?

Name the structures formed from condensation of mesodermal tissue in the area destined to be the pharynx of the developing embryo.

The first pharyngeal arch cartilage, Meckel's cartilage, differentiates into what five bony structures?

Name the muscles that differentiate from the first pharyngeal arch mesoderm?

What cranial nerve supplies the first pharyngeal arch and its derivatives?

Card # 115

- ligamentum venosum

- internal iliac a.

- pharyngeal arches

- incus
  lingula
  malleus
  mandible
  maxilla

- the muscles of mastication: masseter
  lateral and medial pterygoids, temporalis
  anterior belly of the digastric
  mylohyoid
  tensors palate and tympani

- mandibular (V3)

What bony structures differentiate from the second pharyngeal arch cartilage?

Which cranial nerve supplies the second pharyngeal arch and its derivatives?

What parts of the hyoid bone develop from the third pharyngeal arch cartilage?

Which cranial nerve supplies the third pharyngeal arch and its derivatives?

What nerve supplies the fourth pharyngeal arch and its derivatives? The sixth pharyngeal arch?

Which pharyngeal pouch gives rise to the auditory tube and middle ear?

Card # 116

lesser horns and upper part of the body of hyoid bone stapes styloid process	
facial (VII)	
greater horns and the lower part of the body of the hyoid bone	
glossopharyngeal (IX)	
superior laryngeal n. recurrent laryngeal n.	
first	

The thymus is derived from which pharyngeal pouch?

Which pharyngeal arch artery gives rise to the pulmonary artery?

Which pharyngeal arch artery contributes to internal carotid artery?

Which pharyngeal arch artery persists as the right subclavian artery?

Name the interatrial opening in the fetal heart.

In the newborn, blood pressure closes the foramen ovale to form what muscular depression in the heart?

Card # 117

- third (ventral)
- sixth arch a.
- third arch a.
- fourth arch a.
- foramen ovale
- fossa ovalis

Venous blood returning to the heart via the superior vena cava bypasses the lung via what fetal vessel?

Name the fibrous remnant of the ductus arteriosus.

The base of the skull is formed by which type of ossification? The skull cap? The face? The jaw?

What type(s) of ossification form the temporal bone?

Name the three excretory organs that develop in the human embryo. Which one persists in the child?

The ureteric bud gives rise to what tubular structures?

Card # 118

ductus arteriosus	
ligamentum arteriosum	
endochondral membranous membranous membranous	
endochondral: petrous part styloid process membranous: squamous tympanic	
pronephros mesonephros metanephros metanephros	
calyces (major and minor) collecting duct renal pelvis ureter	

The mesonephric or wolffian duct form which male reproductive structures?

What duct gives rise to the uterus and vagina?

What developmental process accounts for the increase in the size of the liver during the second month of gestation?

The pancreas forms by the fusion of what two structures?

The spleen develops from mesenchymal cells of what mesenteric structure?

Which artery supplies the foregut derivatives? The midgut? The hindgut?

Card # 119

- ductus deferens
- efferent ductules
- epididymis
- ejaculatory ducts
- seminal vesicles

---

paramesonephric duct (Mullerian)

---

hemopoiesis

---

dorsal and ventral pancreatic buds

---

dorsal mesogastrium

- celiac a.
- superior mesenteric a.
- inferior mesenteric a.

A deficiency of what substance, produced by type II alveolar cells, leads to respiratory distress syndrome in the newborn?

What depression in the pharynx separates each pharyngeal arch?

Name the term for a large area of separation between the bones of the cranial vault of a newborn that are filled with fibrous tissue.

- surfactant
- pharyngeal pouch
- fontanelle

Distribution of Cutaneous Nerves to Palm and Dorsum of Hand

Illustration 1:1

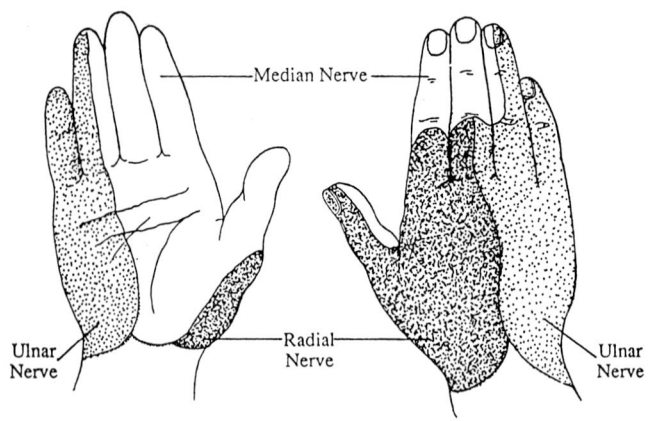

Anterior Thoracic Wall
Illustration 2:1

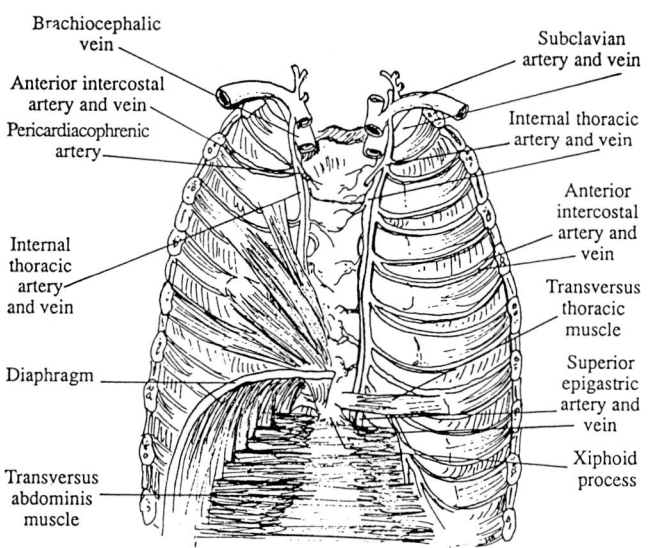

Internal View

Deltopectoral Triangle
Illustration 2:2

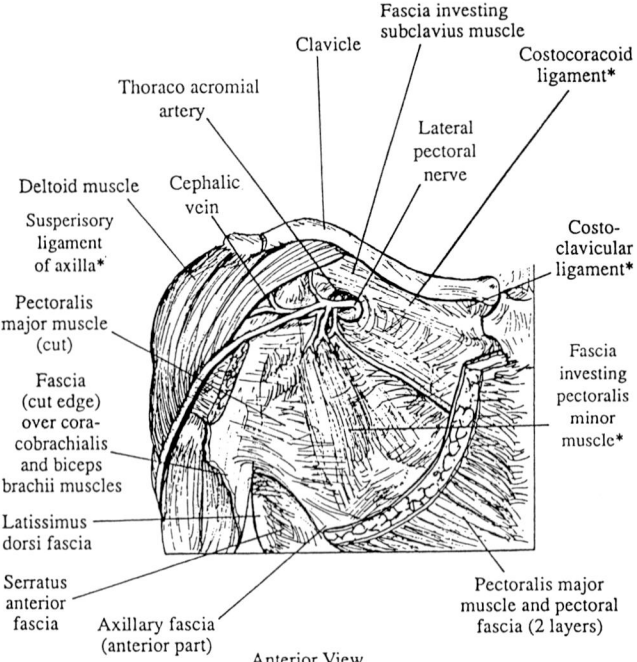

Anterior View

* = Components of clavipectoral fascia

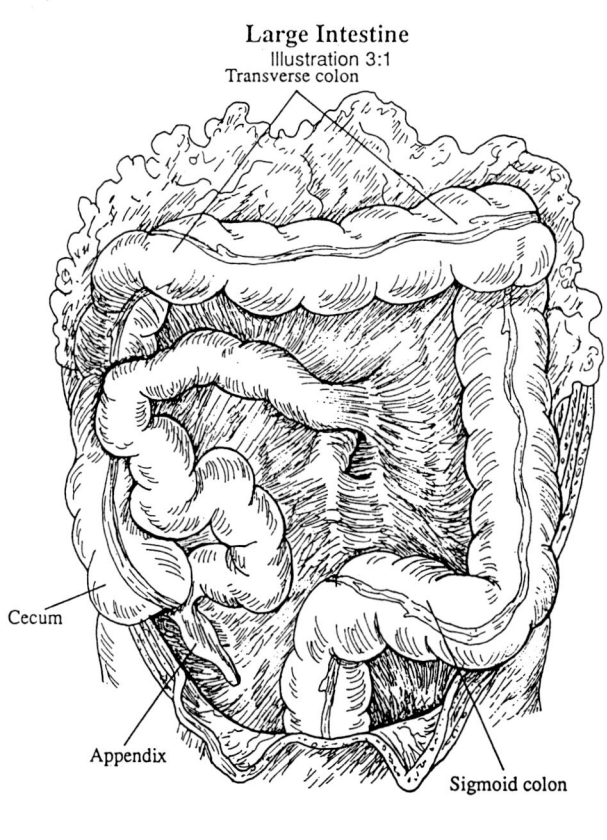

Iliac Arteries And Their Branches

Illustration 4:1

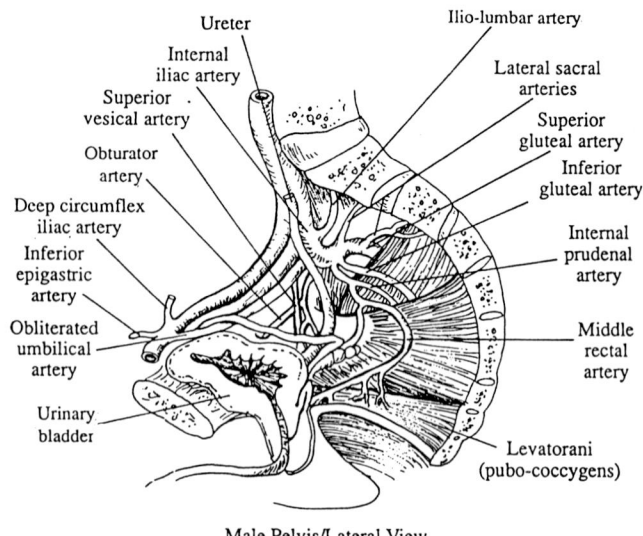

Male Pelvis/Lateral View

Arteries and Nerves of Thigh
Illustration 5:1

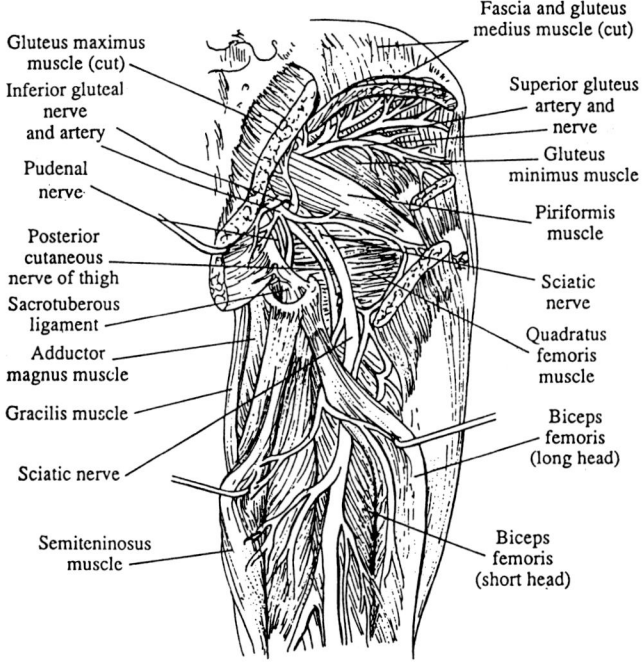

Posterior View

Orbital Cavity
Illustration 7:1

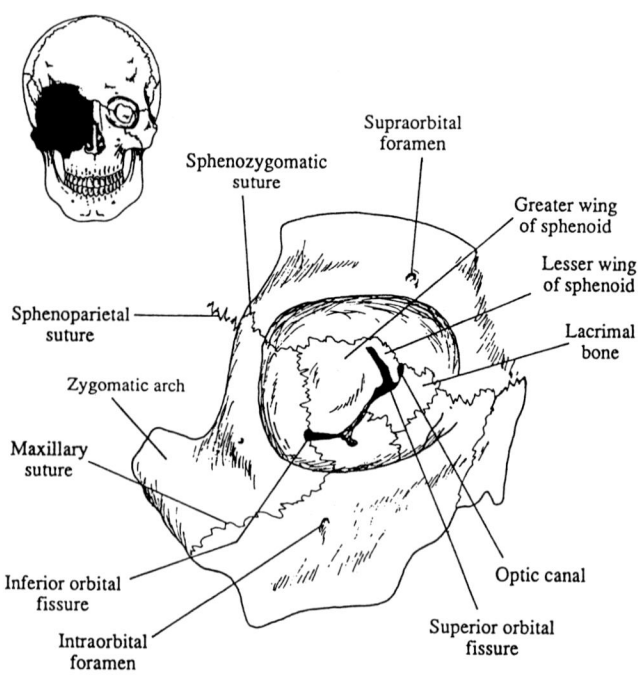

Right Orbit

Laryngeal Cartilages
Illustration 7:2

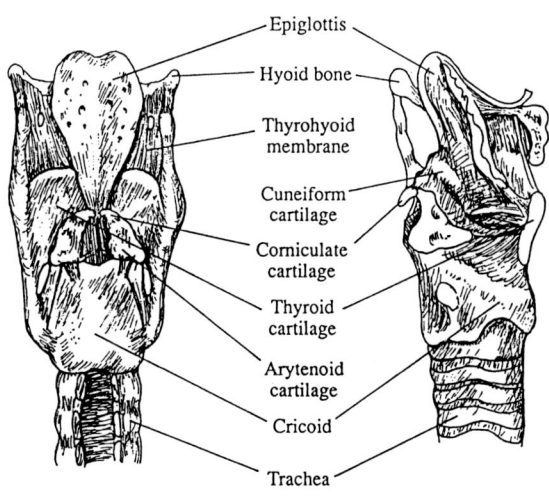

Posterior View   Lateral View